WAGON
CARTS

David Viner

SHIRE PUBLICATIONS

Published in Great Britain in 2011 by Shire Publications Ltd, Midland House, West Way, Botley, Oxford OX2 0PH, United Kingdom.
443 Park Avenue South, New York, NY 10016, USA.

E-mail: shire@shirebooks.co.uk www.shirebooks.co.uk

A CIP catalogue record for this book is available from the British Library.

Shire Library no. 467 • ISBN-13: 978 0 74780 676 9

Designed by Ken Vail Graphic Design, Cambridge, UK and typeset in Perpetua and Gill Sans.
Printed in China through Worldprint Ltd.

11 12 13 14 15 11 10 9 8 7 6 5 4 3 2

COVER IMAGE

The Wagon by Wilmot Pilsbury (1840–1908) shows an old broad-wheeled Sussex wagon (location unidentified). (Reproduced courtesy of J. Collins & Son www.collinsantiques.co.uk)

TITLE PAGE IMAGE

Around the country, wagon and cart makers exhibited local speciality and distinctiveness, their products travelling far and wide.

CONTENTS PAGE IMAGE

The 'Light Essex Pattern Wagon', offered in 1886 in the sale catalogue of Wm Ball & Son of Rothwell. It was built on an oak frame and described as 'a very superior light skeleton Harvest Wagon, being very light in draught'.

DEDICATION

In appreciation of Geraint Jenkins, John Vince and the late David Wray, James Arnold and John Thompson whose illustrations and records have inspired my interest for nearly forty years.

ACKNOWLEDGEMENTS

A great many people, including fellow museum curators, collectors, and horse folk of all kinds, have encouraged my interest in wagons and carts and all are thanked. Such a community shares its knowledge freely, fearful that otherwise it may be lost. Thus are traditional ways maintained and long may it remain so.

Roy Brigden, Diana Zeuner and Catherine Wilson are particularly thanked for their help over a number of years; Freda Gittos and Howard Beard once again gave ready access to their historic postcard collections. My particular debt is to the Museum of English Rural Life, part of the University of Reading, wherein my interest was first aroused, and which has provided study opportunities ever since. A MERL Research Fellowship in 2006–7 was particularly appreciated.

Contributing to the growing Distributed National Collection concept, a National Register of traditional farm wagons and carts is underway and contributions are always welcomed. So too is further information about any of the photographs used in this volume, c/o Shire Publications.

This book seeks to follow in the footsteps of one of Shire Publications' earliest volumes, *Discovering Carts and Wagons* by John Vince (1970), and again makes use of David Wray's splendid illustrations, part of a considerable archive. Photographs and illustrations are from the author's collection except as follows, acknowledged with many thanks, and where copyright is reserved: Museum of English Rural Life, University of Reading pp. 3, 19 upper, 31 middle (copyright F. Foster, Southwold), 35 lower, 37 upper, 39 upper, 42, 44 centre; 45 upper, 46, 54 upper; Lincolnshire County Council (Museum of Lincolnshire Life) pp.31 top left and right; Milton Keynes Museum p.33 upper; St Fagans: National History Museum, pp.36 upper; Weald & Downland Open Air Museum pp. 5 lower, 38 lower; Trustees of Bingham Library Cirecester p.45 lower; Museum of East Anglian Life p.51 upper; Freda Gittos pp.9 centre, 20 upper, 53 lower; Diane Zeuner pp.10, 34 bottom, late Hilary Cotter; Howard Beard pp.17 upper, 19 lower, 48, 52; Catherine Wilson p.18 lower. Where items shown are on display, the museum or site location is given.

CONTENTS

INTRODUCTION

CELEBRATIONS around the country to mark the 'advent of a new Elizabethan era' at the Coronation of Queen Elizabeth II in 1953 included at least one proposal which brought a letter to *The Times* in protest. In the village of South Petherton in Somerset it was suggested, or so a local paper reported, that the event be marked by 'the biggest bonfire ever, to be made among other things from all the obsolete waggons in the parish, thus confirming the passing of the horse and cart in favour of tractor and trailer'. The enthusiastic local committee gave this full support, asking that owners of 'all useless carts and waggons' should take them to the Recreation Field and place them on the bonfire. Thereafter all iron parts left behind would be sold for scrap to benefit Coronation funds. In such a way is history made.

Loading the wagon high. Haytime and harvest were the two key times of the year to bring out and rely upon the farm wagon. Cirencester Park, c.1905.

Left: Jimmy and Joey, reputedly the last two working oxen in England, with the Bathurst Estate's surviving bow wagon, still in its yellow and salmon-pink livery and at virtually the end of its working life. Cirencester Park, *c.* 1960.

The assumption that horse-drawn farming equipment had had its day and was fit for no more than the bonfire sits well with those times. The heyday of the horse as the principal means of power on the land had passed. Once the austerity years in the period immediately following the Second World War had begun to fade, these changes gathered pace so that only the reluctance by some farmers and country people to finally break the link with 'grandfather's old wagon' kept any vehicles intact at all. In function and form they moved from useful, and at one time essential, pieces of farming equipment into a world of nostalgia and memory. Those cast aside at that time have long since rotted away or, indeed, been burnt.

More than half a century later, as the memory fades still further, it is fascinating to discover that numerous wagons and carts do still survive, as collectors' items and as significant objects in the regional agricultural museums set up in a flurry from the 1960s onwards as a record of a way of life being lost from view in any meaningful economic or efficiency sense. What remains in the twenty-first century to the enthusiast, the collector and those who love the 'old country ways' is a scattered nationwide collection, counted now in hundreds rather than thousands.

Museums and collectors hold some of the finest unrestored examples; others have been restored for display in the show ring,

Below: Deep in the shed, this Kent wagon from the Sussex border exhibits the mix of robust design and attention to detail which makes traditional farm vehicles so aesthetically pleasing.

Cart and wagon make a delightful scene in this unidentified location.

as functioning appendages to the healthy interest in the preservation of rare breeds of horses. Very few are used for farming in any real sense and, where they are, horse, wagon, cart and owner are rightly celebrated for their perseverance. Such examples are a link with the past, another category of rare breed, no less than heavy horses, to be treasured and cared for.

The protest letter to *The Times* came from the newly formed Museum of English Rural Life at the University of Reading, then a fledgeling institution appealing for donations and beginning to build up the nationally significant collection it holds today. Its then director suggested creating local exhibitions celebrating the past as a more positive way of marking Coronation year. Rural-life museum curators and other sympathisers have been campaigning similarly ever since, as change after change has revolutionised agricultural processes and equipment across the land.

This book seeks to bring together the story of the variety of cart types which can be found throughout the British Isles, and in particular the regional styles of farm wagon, known by different traditional county names across England and Wales. The fascination of the subject lies not only in such physical variation, and all the reasons why this should have developed at all, but in appreciating the balance of form and function this represents, the sense of traditional craftsmanship – the skills of village wheelwright and, later, larger workshops, and in the idea that such attention to their construction rendered these vehicles almost works of art in the eyes of some beholders. Skills linked to a sense of place and of following in the footsteps of craftsmen who had gone before are key themes.

More colour is added by the wide variety of names given to the various parts of a wagon or cart. An attempt (inevitably incomplete) is made to list them, and also some of the local variants which reveal surviving dialect forms of words around the country. Why are ladders known as 'thripples' in Herefordshire and wagon poles as 'gormers' in Yorkshire? Why is the tailboard

Contrasts in design, shape and ladder form: a Hereford wagon with single-hooped wheels and a Vale of Berkeley wagon with double-straked wheels creating double the width. (Museum of English Rural Life, Reading)

Celebrating the completion of another wagon at this village workshop in the English southern counties (exact location unidentified). The wagon has a shallow body with a gentle fore-end sweep.

of a cart called a 'hind hawk' in Kent, or the mid-rail along the side of a Gloucestershire wagon called a 'dripple'? The answers lie deep in local language, but that the names survive at all indicates the strong sense of tradition which surrounds these vehicles. Even the basic names listed in the Glossary contain many surprises, words which are not used in any other context.

One conundrum stands out – is it 'waggon' or 'wagon'? The well-known writers on this subject have taken their choice. James Arnold had 'waggons', others have 'wagon'; the former seems the older word, and more associated

Pride and joy – posing in the farmyard for the camera. The fore-end of this wagon has both poles and a ladder and the wagon's elegant line is apparent.

Horse team and Sussex wagon in finery for a wedding procession at Ringmer in Sussex.

Gathering the harvest at Newchapel, Surrey.

perhaps with the use of road vehicles rather than in agriculture. But here, as everywhere in this subject, variations are many and often very local. As John Vince wisely said, '... it is not possible to be dogmatic about anything concerned with carts or wagons'. The shorter version is used in this book.

A wain is both a form of cart and a harvest wagon and was once a common name in the West Country; a tumbril is another name for a cart and one of the oldest, often found in East Anglia. But consistently a cart is (or should be) two-wheeled and a wagon four-wheeled.

Most of the wagon examples included here date from the nineteenth and early twentieth centuries, and some from the late eighteenth. Pursuit of the earliest-surviving dated wagon has not progressed further back than 1780.

Wagon head and side boards from Warwickshire owners, stored in Chedham's Yard wheelwright's shop at Wellesbourne.

ORIGINS AND FUNCTION

ANIMALS as well as humans have been the beasts of burden throughout history. Even into early modern times in Britain the packhorse was a vital part of the conveyance of goods around the country, although little more than its high-level routes and the occasional packhorse bridge survive as evidence. It is the development of the wheel in the Middle East from prehistoric times which created an alternative way, allowing simple carts to be constructed, essentially platforms or boxes on wheels, which over time have evolved for aesthetic but essentially functional reasons into an amazing world-wide variety of types, shapes and forms.

The cart as well as the sledge has a long pedigree in the British Isles. Farm vehicles in medieval times were usually two-wheeled ox- or horse-drawn carts, whilst in hilly areas use was made of the sledge or slide-car. Four-wheeled wagons were usually for transport by road of people, goods and baggage. Medieval documents such as the Luttrell Psalter, dating from the mid-fourteenth century, show examples of both. One document dated to 1496 depicts a four-wheeled wagon fully laden with merchandise.

From the thirteenth century, particularly in the south and east, there was an increase in the use of horses for traction but it was not until the seventeenth century with the technical development of the movable fore-carriage or pivoted front axle that the four-wheeled wagon became a useful vehicle for farm work.

The Psalter provided the stimulus for the modern re-creation of a typical all-purpose cart of the fifteenth century as part of an interpretation of buildings of similar date re-erected at the Weald & Downland Open Air Museum in West Sussex. Oak, ash and elm were used in the construction and, on an all-wooden axle, the wheels were shod with sections of iron strakes rather than single iron hoops. The simple platform resting on the axle and the open spindle-sided frames create a very light, manoeuvrable vehicle. This cart remains one of the few re-creations of its kind, although in shape and design its like can still be seen in use in Eastern Europe today.

Opposite: A replica medieval cart displayed outside the early-fifteenth-century Bayleaf farmhouse at Weald & Downland Open Air Museum. Between the shafts is a Dales pony of a size typical for the period.

THE WHEEL

The Glossary lists a mass of special names for the individual parts of wagons and carts; in addition, regional variations in naming abounded. It is quickly possible to become confused! However, the wheel is clearly central to the safety and efficiency of any cart or wagon.

It consists of a series of spokes, usually made of oak for strength, carefully but firmly tenoned into the central hub (also known as a nave or stock). The rim is created from a series of felloes into which the spokes are fixed, and the whole is secured by an iron banding, applied under heat so that it shrinks to secure a tight fit. An alternative was a series of strakes, which were nailed into position overlapping felloe joints around the wheel. Wheels of this latter construction are often taken to indicate an older type and date of vehicle. Different types of wood were used: always elm for the hub and ash or beech for the felloes.

The shape of the wheel in the vertical plane is important too. It is said to be 'dished' when set at an angle to the main body as a means of counteracting the sideways pressure of the load on a vehicle and dealing with the swaying movement of the horse. So a moving wagon or cart can seem ungainly but is in fact well balanced. Around the country the degree of dishing varied, and greater dishing, like strakes and other features, is often taken to indicate greater age. This may be so, but is rather more an indication of functionality in the countryside where vehicles were required to work hard.

Creating a wheel to achieve all these things is the work of a skilled craftsman.

Right:
This wheel exhibits all the characteristics of a heavy-duty 4 foot diameter wheel from a Border Counties cart: the broad felloes are built to accommodate a wheel 6 inches wide and made from a pair of strakes, each nailed into place. The wooden axle has a large hub.

Far right:
The all-wooden axle, nave (or stock) and lynch pin on the medieval replica cart, supported by two iron stock bonds.

THE CART

THE TWO-WHEELED CART, on a single axle and drawn by ox, donkey or a
of work on the land for
al for short-distance farm
it remained so until the

used and more goods or
Britain today date from
nwards. Unlike a wagon,
mble cart was worked to
t few if any examples of
storical objects.

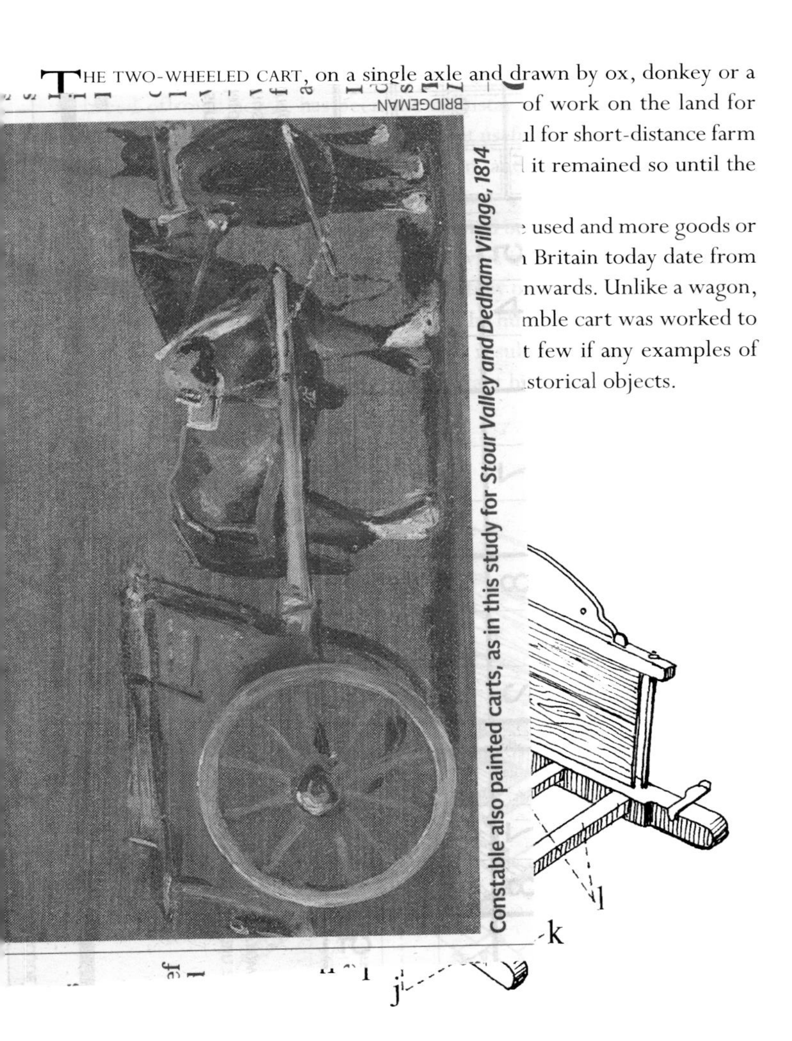

Constable also painted carts, as in this study for *Stour Valley and Dedham Village, 1814*

A Hertfordshire cart, showing the typical build of a dung cart:
(a) forehead
(b) top rave
(c) top board
(d) shaft
(e) strouter
(f) dirt board
(g) iron axle arm
(h) clip
(i) axle or ex bed
(j) tail ladder
(k) side or sole
(l) shutters.

A restored cart in use at the National Folk Museum at Cregneash, Isle of Man.

One such is certainly the surviving body of an ox wain or cart from a farm at Ewenny in the Vale of Glamorgan, built between 1750 and 1770 by Thomas Thomas, a Glamorgan wheelwright whose initials appear on the front headboard. The cart is light and heavily chamfered and its chief characteristics are the spindles mortised into the frame and the distinctive arch of both inner and outer top-rail, strongly resembling those of the larger bow wagons also made in Glamorgan. Its long boarding is often taken as another indicator of age although this (like strakes) may be more functional than chronological in significance.

Writing about Herefordshire in the 1790s, William Marshall noted the local use of the word 'wain' for an ox cart, and commented that as recently as fifty years before 'the wain was the only farm carriage of the district, there being many men now living who remember the first introduction of waggons'. Here is evidence of the larger vehicles being added to the farmer's stock, but not replacing the humble workaday cart with its single horse as the basic tool for many jobs.

The distinctive open frame of an ox cart from Ewenny, Vale of Glamorgan, built between 1750 and 1770. (St Fagans: National History Museum)

Oxen were used in Sussex into the early twentieth century. This pair, believed to be near Lewes, are harnessed by pole to a simple, heavily framed cart.

Farmyard scene at Huxley Farm, Hedge Lane, in 1930.

TYPES

There are a number of distinct cart types, based as much on functionality as regional variation. Unlike wagons, their story covers the whole of the British Isles, so that local forms can be studied in Wales, on the Isle of Man, across northern England, and in Scotland and Ireland. Terminology varies too, as might be expected.

The basic box construction often had the simple role of transporting dung to the fields, hence the tipping mechanism of some of them. Size again depended upon the demands of the landscape in which a cart worked: heavier for example for bigger loads in open East Anglian countryside than in the narrow lanes of Devon.

Heavy loads such as dung and also root crops or stones put pressure on the cart sides, so a heavy, well-supported structure with large

A plank-sided cart, simple in line and robust in construction, used by 'Silvester, contractor'.

The owner's name on a Lancashire cart.

'strouters' was required. Sides were planked or panelled. Chamfering is another feature sometimes found, with the advantage of reducing unnecessary weight. The basic frame or bed with long and heavy sides could have summers running longways; alternatively the frame was cross-timbered with shutters. It was important to provide a framework for the long boarding which had a practical purpose when shovelling loads from the vehicle. The capacity of a cart could be increased by fitting extra boards to the sides and front to create more height. The average cart could carry a load of from 18 to 22 cwt.

Carts with no tipping mechanism were cumbersome to unload, the horse

Reality, routine and the regular use of the muck cart.

A comparison in styles: (*left*) a panel-sided cart on iron hubs by Bingham & Son, Makers, Long Sutton in Lincolnshire, dated 1921; (*right*) a heavier Border Counties cart on double strakes. Brongain Farm collection sale, Michaelmas 2007.

having to be released from the shafts. Methods of tipping varied, pivoting the cart body on the shafts or the axle, and provide interesting details to spot in surviving examples. A simple chain might control the degree of tip; there were various screw-type options and others such as a bar, a strap stick, a lever, a sword and lever, and a sword and peg. The sword or 'tip stick' is a vertical fitting, usually of iron, controlling the angle; sometimes makers' names can be found on them.

At harvest time, the length of a cart could be extended by the use of ladders in a variety of ways, either upright or sloping at a steep angle to contain a load, or laid low and forward from the front to extend the

This restored hermaphrodite cart/wagon clearly shows the length of the harvest platform once the ladders are fully extended.

platform. Care was required in loading so as to keep a balance and not to put too much weight on the horse.

A half-wagon, half-cart type called a hermaphrodite was a particular feature of the East Midlands and East Anglia, where it was known as a 'Moffrey' and 'Morphrey' respectively. The body of the cart was turned into a four-wheeled wagon for the harvest season by the addition of a front axle, this being stored away for the rest of the year.

This Scotch cart bears the farmer's name: 'Thos Pattison, Hall Farm, Seaton Delaval'. It is seen here at a collection dispersal in Norfolk. Its then owner's family had moved from Scotland to East Anglia in the late 1920s.

Farm staff on a long cart at Home Farm near Newcastle with George Summers, farm steward, on the right.

MANUFACTURE

Carts were locally made in village or town workshops, and demanded skills from the wheelwright, blacksmith and carpenter, often one and the same. Mass production of carts dates from early in the nineteenth century; one feature was the use of cast-iron hubs for wheels showing the maker's name as both advertisement and a recommendation of quality. These same processes of adaptation in construction methods are seen in both carts and wagons.

As the nineteenth century progressed, carts began to be imported from Scotland into East Anglia. The term 'Scotch cart', originally specifically for these, became more widely used for many factory-made types to the same or similar design. North of the border, cart production was usually the joiner's stock-in-trade and certain makers such as Jack & Sons of Maybole, in Ayrshire, became well known. Many carts came with their owners, who moved south to farm in the eastern counties of England. Not surprisingly, East Anglian makers gained a reputation for building Scotch carts.

A pair of harvest carts break their journey in a village street.

The term 'Scotch cart' is also used for harvest carts, long-bodied and not designed to tip, which are another important category of cart, particularly in the west of

Haymaking in West Wales, with a local type of long cart. This postcard was sent from Llanstephan in 1905.

England, the northern counties and the Scottish border. The links between these and the bow or 'hoop-raved' design will be explored later (p. 43). This basic shape, but capable of being tipped, was also marketed in both a simple plain-sided and more elaborate spindle-sided forms by companies such as Bristol Wagon & Carriage Works Co. Ltd, who claimed that it could carry 30 cwt.

These apart, while there are strong regional and traditional variations in long carts, they are characterised by a long side body, often with integral shafts, and the minimum of impediment to the loading platform. The form of protection for the wheels is often the diagnostic feature, from the simple hoop rave to various forms of ladder. Devon has a ladder-sided form known as a 'curry cart' and there are numerous Welsh variations such as 'gambo' and 'longbodies'.

When studying carts, it must be remembered that vehicles of such simple design derive from the basic sledge, which remained into the twentieth century a favourite form on the steep hillsides of the uplands, except in Radnorshire where a particular form of wheelcar was used.

A 'peat cart' or sledge used by the Anderson family until the 1920s at Stone House, Keld, Swaledale. (Dales Countryside Museum, Hawes, North Yorkshire)

THE WAGON

IN ORIGIN four-wheeled wagons are also of simple construction, little more than the joining of two carts together, with the same open spindle-sided form. Wattles or hurdles could be added whenever loose goods were to be carried. The use of a coupling pole to make the connection is simple but effective, with a pivot on the front axle, and in this standard form the wagon stood alongside the cart as a basic conveyance of crops and materials on the land for generations. In use on the roads, it was always subject to the poor state of the highways, and this inhibited development and structural improvements until the roads themselves came to be improved in the early post-medieval period.

This chapter introduces the enormous variation in design for the body of four-wheeled wagons, looking at their different shapes, whether box, bow or more standardised later designs, as wagons evolved from their medieval precursors. On the other hand, as with carts, wagon bodies sit upon an undercarriage or frame essentially the same in shape and function across the range. When looking at a wagon or cart today, however, it is important to remember that the undercarriage and the body may not be of the same age, one or both having been replaced over time, as indeed may any one of the wheels.

BASIC STRUCTURE

The coupling pole is the central device linking front and rear axles. It is braced and fixed to the rear axle, where both pole and braces rest upon the axle bed;

David and Goliath – on the left a Dorset box wagon made in Broadoak, c.1900, and on the right an East Anglian wagon used at Hadleigh in Suffolk, standing 6 feet high at its mid-point. (Museum of English Rural Life, Reading)

Variety in style – the front boards and waists of four very distinctive regional designs. Raves (top, outer and mid-rail) are much in evidence; so too strouters, standards and spindles supporting and decorating the wagon sides.

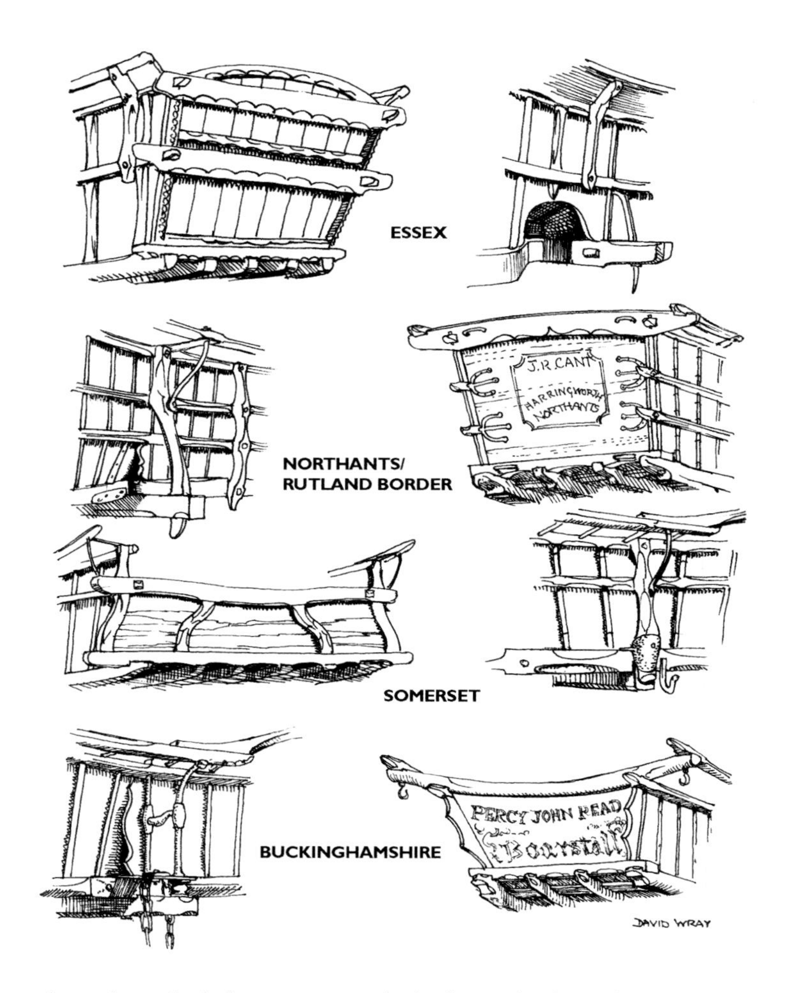

above them the bolster supports the body. At the front the wheels are generally smaller, not only for reasons of manoeuvrability but also to create space for the extra layers of timber required to form the fore-carriage. The fore-carriage includes the pillow and the hounds, the means by which the front axle is kept in vertical alignment, preventing it from buckling forward or backwards against the strains of movement.

LOCKING

The front axle controls the degree of turn or 'lock' which a wagon can achieve. Each of the regional types discussed below have different degrees of lock, reflected both in the shape of the frame at the mid-point where the

The front undercarriage of a large East Anglian wagon: from the bottom upwards, the coupling pole sits upon the axle, balanced by the hounds and the slider bar, all beneath the slightly arched bolster, above which sits the large pillow, with a pair of distinctive open eyes for both decoration and weight reduction. (Museum of English Rural Life, Reading)

turning front wheels meet the body and in the size and positioning of the fore wheels. One of the advantages of dishing is that it creates extra space between wheel and body, thus increasing turning capacity.

There are four categories of wagon lock, from the least to the most flexible. First, the quarter-lock wagon has a straight-sided body without indentation, so that the wheels can only turn until they strike the body, where

Detailing of the waisting on a Rutland wagon, showing both the inset, the distinctive crossledge and wooden strouter, and the pair of middle rails with the owner's nameboard set between them in typical fashion. (Rutland County Museum, Oakham)

iron plates provide protection. This type also has large front wheels. Some regional types such as Staffordshire or Surrey were particularly associated with this form. It is amongst the heaviest to use and the least easy to manoeuvre and, obviously, needs a large turning circle.

Half-lock wagons have a waisted body, set just in front of the crossledge and central strouter. Thus the wheel can gain a greater turn into the wheel arch, decreasing the lock required. Many of the regional types exhibit this feature – sometimes it is quite pronounced – and it is fairly common. Both quarter- and half-lock wagons formed the mainstay of village workshop production, the later and more manoeuvrable types (as below) developing in the period of factory production, although it would be wrong to assume that village workshops did not also move with the times. Equally, in certain areas such as Gloucestershire techniques such as straking often persisted until the very end of regional wagon making.

Three-quarter-lock wagons have smaller fore wheels which can be turned below the wagon bed, if only as far as the coupling pole. Their flexibility also made them popular as road vehicles for carriers and other regular road users.

Finally, fully locking wagons had no restriction for the fore-carriage, as there was no pole and the wheels could turn at right angles under the body. The balance for the fore-axle was achieved by a horizontal fifth wheel, a strong iron ring fixed to the wagon bed and the turning bolster. The smaller front wheels, noticeably smaller than the other examples, completed the assemblage.

Such wagon locking forms were not simply consecutive in development. Much variation and overlapping can be seen when studying surviving wagons and it is a mistake to date locking variations in strictly chronological order. The use of iron rather than wood for side supports and for fixings holding the outer raves in place is another variable that is hard to categorise, as such items were often bought in ready made by wagon workshops as they developed during the nineteenth century.

EVOLUTION

It is generally argued that the more extensive development of four-wheeled wagons throughout Britain, and especially the evolution in the variety of forms, dates from the campaigns to drain the land in the eastern counties of England from the sixteenth century onwards. The similarity between the narrow-bodied and high-sided box wagons of the Low Countries and the wagons of Lincolnshire strengthens this view. Variations of this distinctive type can be traced all across the East Midlands and East Anglia in the following centuries.

A key issue throughout the story of wagon evolution is the relationship between this box form and the shapely bow or hoop-raved wagon design. Box wagons, although seemingly uniform, show marked local variations across the whole of central, southern and south-western Britain from at least the eighteenth century onwards. The distinctive bow or hoop-raved shape is much associated with the South Midlands, and in its finest form is named after this region. It is also a strong feature in the south-western counties and South Wales, suggesting a different tradition perhaps directly related and evolving from a type of two-wheeled cart much used in this area.

This relationship between box and bow and the interplay between cart and wagon design have occupied many writers and led to a range of conclusions. Wagons vary hugely across the regions and agricultural vehicles in particular show endless adaptations. This makes virtually every surviving locally made wagon potentially a unique example of its kind.

Identifying the wagon or cart maker is not always easy. Clues on axle hub-caps are not always reliable; side and front plates in various designs and material provide better clues.

ON THE ROAD

Apart from agricultural purposes, another key strand in the development of wagons was the use of heavy vehicles for road haulage by a network of long-distance carriers. Until the coming of the railways, this was the way goods were moved around the country. The routes led mostly from the countryside and rural towns into the cities. Because of the poor condition of the roads, a mass of legislation arose from the sixteenth century onwards. This, however, dealt with the problem simply by controlling the use of the roads. Only with the system of turnpike trusts, raising money locally and charging for the use of individual routes, do any real improvements, especially in road surfaces, begin to take effect.

The enormous lumbering stage wagons, running fixed routes for the conveyance of a mass of goods, began to benefit from these improvements from the later seventeenth century onwards. This is reflected in their overall design. They became less bulky and less reliant upon a large team of horses, sometimes six to eight in number, to haul them through the mud.

The regimes imposed on such vehicles were based upon potential for damage, and so a series of wide wheel Acts from 1662 onwards penalised the use of narrow wheels

William Pyne's representation very early in the nineteenth century of a fully laden stage wagon, lumbering past the forty-mile milestone with an eight-horse team and no fewer than five strakes to each wheel.

on the grounds that they were more damaging to the road surface. Wide wheels were encouraged, and this is particularly noticeable in illustrations of stage wagons, where several rings of strakes on pronounced dished wheels can be seen.

It is believed that none of the wide-wheeled types such as illustrated by Pyne have survived, although the Museum of English Rural Life at Reading displays what is probably the oldest example of a road wagon, believed to date from 1780 and made at Horseheath in Cambridgeshire. Its wheels now have only 4 inch hoops, still robust enough, but much of its detailing looks old, in a style which can be traced through many other wagons of later date from this part of the south-eastern Midlands. Many farm wagons were also used for road work, of course, and this one was owned and used by Mr Webb of Streetley End to carry goods between there, Cambridge and London, a journey taking four days in each direction. It was amongst hundreds of others coming into the capital from rural areas.

The simple box form of this late-eighteenth-century road wagon from Cambridgeshire shows clearly in this view; long boarding survives and there is little impediment to packing large quantities of mixed goods for long-distance carriage, under a canvas cover. (Museum of English Rural Life, Reading)

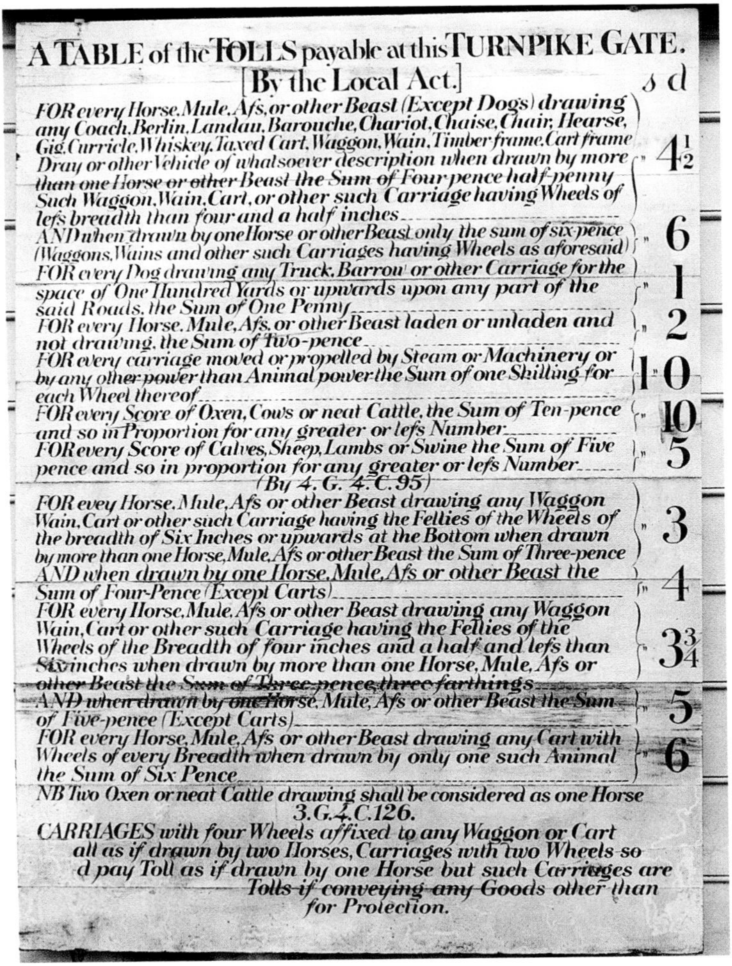

A TABLE of the TOLLS payable at this TURNPIKE GATE.
[By the Local Act.]

	s	d
FOR every Horse, Mule, Afs, or other Beast (Except Dogs) drawing any Coach, Berlin, Landau, Barouche, Chariot, Chaise, Chair, Hearse, Gig, Curricle, Whiskey, Taxed Cart, Waggon, Wain, Timber frame, Cart frame Dray or other Vehicle of whatsoever description when drawn by more than one Horse or other Beast the Sum of Four pence half-penny Such Waggon, Wain, Cart, or other such Carriage having Wheels of lefs breadth than four and a half inches	"	4½
AND when drawn by one Horse or other Beast only the sum of six pence (Waggons, Wains and other such Carriages having Wheels as aforesaid)	"	6
FOR every Dog drawing any Truck, Barrow or other Carriage for the space of One Hundred Yards or upwards upon any part of the said Roads, the Sum of One Penny	"	1
FOR every Horse, Mule, Afs, or other Beast laden or unladen and not drawing, the Sum of Two-pence	"	2
FOR every carriage moved or propelled by Steam or Machinery or by any other power than Animal power the Sum of one Shilling for each Wheel thereof	1	0
FOR every Score of Oxen, Cows or neat Cattle, the Sum of Ten-pence and so in Proportion for any greater or lefs Number	"	10
FOR every Score of Calves, Sheep, Lambs or Swine the Sum of Five pence and so in proportion for any greater or lefs Number	"	5

(By 4. G. 4. C. 95)

	s	d
FOR evey Horse, Mule, Afs or other Beast drawing any Waggon Wain, Cart or other such Carriage having the Fellies of the Wheels of the breadth of Six Inches or upwards at the Bottom when drawn by more than one Horse, Mule, Afs or other Beast the Sum of Three-pence	"	3
AND when drawn by one Horse, Mule, Afs or other Beast the Sum of Four-Pence (Except Carts)	"	4
FOR every Horse, Mule, Afs or other Beast drawing any Waggon Wain, Cart or other such Carriage having the Fellies of the Wheels of the Breadth of four inches and a half and lefs than Six inches when drawn by more than one Horse, Mule, Afs or other Beast the Sum of Three pence three farthings	"	3¾
AND when drawn by one Horse, Mule, Afs or other Beast the Sum of Five-pence (Except Carts)	"	5
FOR every Horse, Mule, Afs or other Beast drawing any Cart with Wheels of every Breadth when drawn by only one such Animal the Sum of Six Pence	"	6

NB Two Oxen or neat Cattle drawing shall be considered as one Horse
3. G. 4. C. 126.
CARRIAGES with four Wheels affixed to any Waggon or Cart ... all as if drawn by two Horses, Carriages with two Wheels so ... d pay Toll as if drawn by one Horse but such Carriages are ... Tolls if conveying any Goods other than for Protection.

A mass of tolls facing the road user at the Northchapel toll gate in West Sussex. Wagons and other vehicles with wheels less than 4½ inches wide paid 4½ pence; those with wheels of 6 inches or over paid only 3 pence. Between the two dimensions the charge was 3¾ pence. (Weald & Downland Open Air Museum)

One feature of the management of roads was the requirement for identification. A notice issued about 1820 by the trustees of the extensive Bath Turnpike Roads Trust required the owners 'of any Waggon, Cart or other Carriage of Burthen' travelling on the Trust's roads 'without his or her Christian and Surname, and Place of Abode, being painted thereon, in large legible Letters, on the Front or Off-Side of the same' to be liable to a fine according to Act of Parliament. During the Second World War a security restriction required owners to disguise the address element on their wagons, hence the occasional blacked-out section still visible on some wagons today.

S. E. MILES & SON.
MELBURY MILLS
SHAFTESBURY.

BOX WAGONS

TURNING to look in more detail at specific regional types, the logical place to start is with the box wagons of the eastern counties from Lincolnshire down through East Anglia, which exhibit a number of common features. These wagons are among the largest in the country, often with the highest bed from the ground (making them not the easiest to use) and the largest wheels; but they were also capable of taking the greatest loads and over longer distances in the open landscape to which they belonged.

The 'Lincolnshire' high-sided wagon is set high above the axles, its spindle design strikingly angled forward and back from the vertical at the mid-point. There is a single mid-rail but often a pair, and without any outraves additional height is gained by the fixing of boards into staples on the top rave, an essential extra requirement at harvest time. Blue was a favourite wagon body colour and the owner's name formed an elaborate decoration on the front headboard. Like all such county-named types, this wagon was not exclusive to Lincolnshire and could be found over several counties. Equally, as with others, its heartland was the county of its name.

Variations in East Anglia can be detected across Norfolk, Suffolk and Essex. There is the same if less pronounced sweep to the top-rail and size is again very apparent, rear wheels of up to 6 feet in diameter being common. There is often quite a deep waisting, and a particular characteristic is the way in which both front and tail boards were detachable, often in upper and lower sections. The strong wooden strouters were supplemented above the mid-rail by others at half height supporting the upper side of the body. In Suffolk particularly, a common variation is to find this upper section open, which creates an interesting half-panelled effect. Many factory-made wagons exhibit this form and even the shallower half-bed type. The flexibility so created made for a very useful all-purpose vehicle. As with so many other wagon types, the favourite body colour was blue, with the undercarriage varying from a light salmon pink into a deeper red.

From South Lincolnshire across much of the Midland counties a recognisable type of all-purpose box wagon can be found, with interesting

Opposite:
The graceful lines and colour scheme of a Dorset box wagon, displayed here at Dyrham Park in South Gloucestershire. Its owner (S. E. Miles & Son, Melbury Mills, Shaftesbury), date (1910), workshop (Birdbush Wagon Works, Donhead St Mary), and even wheelwright (W. M. Burden) are all known.

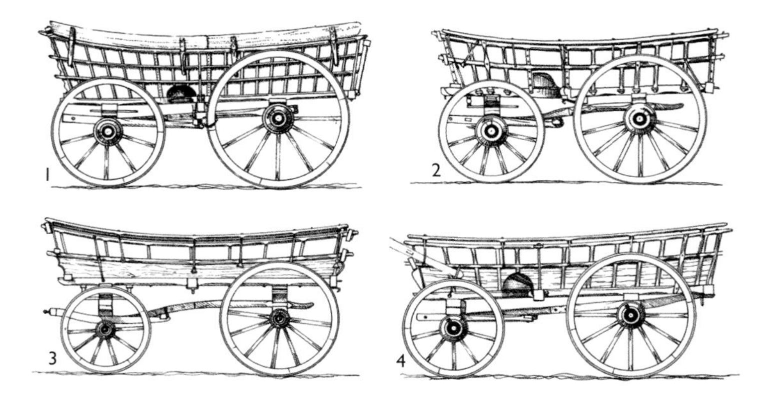

Box wagons from the eastern counties of England: (1) Lincolnshire *c.*1860, (2) Norfolk *c.*1840, (3) Norfolk 'factory' wagon *c.*1920, (4) Suffolk *c.*1880.

variations from place to place. It is called variously a 'South Lincolnshire' or 'Rutland' type, reflecting its core area, which also includes Northamptonshire, part of Leicestershire and southwards into Huntingdon and beyond. Indeed as this box type spread westwards, assuming that this was the flow of its original influence, good examples of all dates can be found in Warwickshire, Worcestershire and north Gloucestershire.

Typically some 10 to 12 feet in length, the body of this type of wagon has outraves, with a noticeable overhang of the top-rail into which a shallow pair of ladders may be fitted as required. A pair of mid-rails, between which is usually the owner's nameboard, have strong wooden strouters supporting the close spindle sides, the central one on the crossledge resembling a

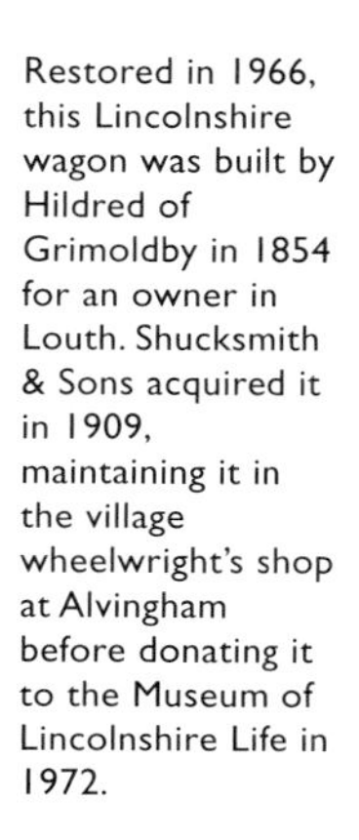

Restored in 1966, this Lincolnshire wagon was built by Hildred of Grimoldby in 1854 for an owner in Louth. Shucksmith & Sons acquired it in 1909, maintaining it in the village wheelwright's shop at Alvingham before donating it to the Museum of Lincolnshire Life in 1972.

Top left: Threshing on James Auckland's farm at Martin Fen in Lincolnshire. The nameboard on the wagon is dated to 1880.

Top right: Sunday School outing at Bassingham, Lincolnshire. In time-honoured fashion, local farmer Joseph Brocklebank loans his box wagon for the occasion. Floral designs adorn both horse and wagon.

Middle: A Suffolk wagon seen here in 1937 on Southwold marshes. The body sweep, high rear and lower front wheels, and detachable half-tailboard are all apparent.

Above: The removable headboard panels on this East Anglian wagon are finished off with intricate chamfering. Used at Paradise Farm, Worlingworth in Suffolk from c.1906 until 1958, it is now in the Museum of East Anglian Life, Stowmarket.

Box wagons from the East and West Midlands: (1) South Lincolnshire *c.*1870, (2) Northamptonshire *c.*1890, (3) Warwickshire 1912.

buttress in shape. The overall impression is of a robust workhorse, with an otherwise uncluttered curving single front headboard decorated with the owner's name and often also that of the builder and a date. The Northamptonshire wagon in particular was renowned for the decorative detailing on the front panel, with a roundel frequently its central feature.

A common colour was an all-over orange, known as 'Rutland red', which certainly made these wagons stand out. Others were in blue or a plain ochre

A South Lincolnshire wagon, repainted in 1976 when in a private collection in the county.

One of the few Midland counties box wagons actually recorded as being from a specific Northamptonshire address: Thomas Messenger of Adstone. It was removed in the early 1970s to the developing Stacey Hill Collection, now Milton Keynes Museum, but was, alas, destroyed in a serious fire in the museum store on New Year's Day 1996.

or stone colour. One interesting feature was the design of some tailboards, constructed to serve as a useful ladder for access into the wagon.

One of the oldest examples of the Rutland wagon survives from the late eighteenth century and is suitably in the County Museum at Oakham. Its faded front roundel reads 'Thomas W. Lawrence, Preston, Rutland' and the wagon is reputed to have been bought second hand by the donor's grandfather at a sale near Gainsborough in the 1790s. Although repaired more than once, its long 12-foot body shows its age. Before its present orange colour, it was painted blue.

Rear tailboard in the form of a convenient ladder. (Coggeshall Grange Barn, Essex).

Although metalwork has replaced earlier wooden features, the pair of mid-rails and close spindle design are strong indicators on this early Rutland wagon of *c.*1780, which also shows a modest waisting. (Rutland County Museum, Oakham)

Closer to London there are further variations in the basic box design. Wagons from counties such as Cambridgeshire, Hertfordshire and Buckinghamshire show distinctive features usually related to the additional role of many farm wagons here as vehicles for the conveyance of goods into the capital by road. The example from Horseheath has already been noted (p. 26). Often shorter, no more than 10 feet, such wagons might also have iron rather than wooden spindles, but they retain the two mid-rails and have narrow wheels no more than 3 inches wide. The body colour might be buff or the all-purpose favourite, blue.

Other than outlining, the only decoration on the front headboard of this Midlands box wagon used at Manton in Rutland is its date of 1854.

Moving westwards, the wagons gathered into county-based collections in Oxfordshire, Warwickshire, Worcestershire and beyond retain the mid-rail feature, but often only a single rail placed higher up than halfway. In some examples this also forms a detail across the front headboard too, distinctively so in and around west Oxfordshire. Others, with examples west of the Severn and into Monmouthshire, were very plain by comparison. Both Thomas Hennell and James Arnold (see Further Reading) recognised a regional type, where the sweep of the body is accentuated by the long ladders fixed in the very centre of the wagon, above one or three dominant strouters.

Agricultural turnouts at shows around the country keep alive a tradition of pride in wagon and cart variations and makers. This Aylesbury road wagon built by J. Plater of Haddenham, Buckinghamshire, is shown by Gwylim Evans.

Gloucester box wagon at the gates of Batsford Park in the north Cotswolds, probably belonging to the estate and recorded by the Hon. David Mitford, a keen photographer.

Otherwise the body is plain boarded, save only for the long nameboard on the side. Double-straked wheels front and rear are a feature, and Hennell also showed the use of harvest poles in addition to the permanently fixed ladders.

Along the Welsh border, in those counties which lead into or are part of the Upper Severn plain, is a richness of wagon design. Beyond into the Welsh hills was cart country, so here is a striking contrast in every way. Detailing is often heavy, as are the wagons themselves, but amongst the wagons are some of the most attractive in the British Isles.

Radnorshire wagons had various forms of harvest work to do, such as taking heavy loads of apples from the orchards. So a strong U-shaped strouter is a feature supporting the body, and standards are robust. Probably the finest surviving county example has been in the national collection in Cardiff since

James Arnold's typically accurate and detailed record of a Hereford wagon, painted in 1965.

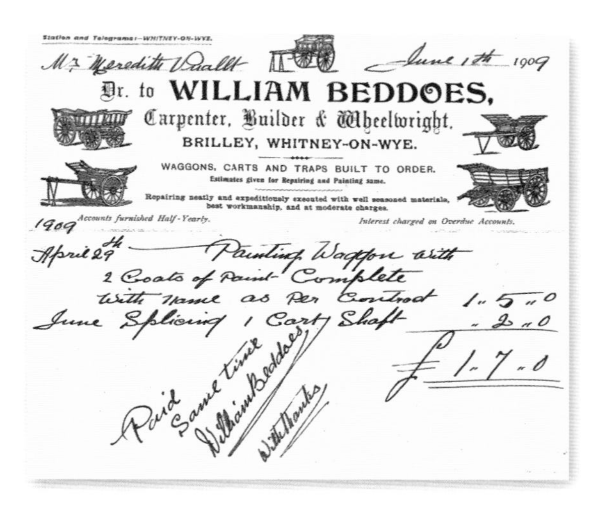
Station and Telegrams:—WHITNEY-ON-WYE.

Mr Meredith Vaallt   June 1th 1909

Dr. to WILLIAM BEDDOES,

Carpenter, Builder & Wheelwright,

BRILLEY, WHITNEY-ON-WYE.

WAGGONS, CARTS AND TRAPS BUILT TO ORDER.

Estimates given for Repairing and Painting same.

Repairing neatly and expeditiously executed with well seasoned materials, best workmanship, and at moderate charges.

1909 Accounts furnished Half-Yearly.   Interest charged on Overdue Accounts.

April 29th — Painting Waggon with 2 Coats of Paint Complete With name as Per Contract 1„5„0

June Splicing 1 Cart Shaft „2„0

£1„7„0

Paid Same time William Beddoes With thanks

1948; it dates from 1897, was made by William Beddoes of Brilley near Hereford, took thirteen weeks to build and cost £35 10s. The wagon's first use was as part of Queen Victoria's Jubilee celebrations in that year.

In Shropshire and Montgomeryshire, a single mid-rail, heavy standards and in some cases a pronounced curve to the profile are all distinguishing features; so too the colour (yellow) and the dishing of broad wheels. Not all were as heavily made and there are the usual variations but they stand comparison with other heavyweight wagons from this region, classified as 'Staffordshire' and 'Denbighshire'. The former county has very few examples surviving. Quarter locking, with a flat profile to the body, closely spindle sided and with a single mid-rail on sides and across the front, and often 12 feet long, they epitomise both strength and weight, requiring several horses to pull when fully laden.

Denbighshire wagons are very distinctive, with details both elaborate and attractive. There is a distinct forward slope to the body and as in Radnorshire these wagons were beasts of general burden. The two mid-rails are evenly spaced on both sides and across the headboard which is busy with

Shropshire wagon c.1915, made by Mr Cadwallader at Bishop's Castle and used nearby at Lydbury North. Compare its size with that of the boat wagon alongside. (Museum of English Rural Life, Reading)

As heavy as they come: this Staffordshire wagon was built c.1860, is over 13 feet long, stands 6 feet high, and has broad wheels with double strakes totalling 6½ inches. (Staffordshire County Museum)

solid standards, the front board being topped off with an arched panel filled with small spindles, reminiscent as we shall see of the Glamorgan style. Colour is often orange or blue, the whole a most attractive vehicle.

In the south-east of England the box wagons of Sussex and Kent form another distinctive grouping, lighter in construction than others so far discussed and generally half locking. A blue body and a single mid-rail were characteristics and so too the idea of a loose body, where the fixings between body and undercarriage differ from other types. The antecedents for this may well lie in the structure of timber carriages used for hauling felled timber – essentially two axles linked together in simple fashion via a long pole. The older broad-wheel Sussex wagon has long been a rarity; researching as long ago as 1936, J. B. Passmore at Reading University, who made a detailed study of wagon types, seemed too late to get a record. His agricultural respondent at West Sussex County Council had not seen one since 1921 and 'everyone

Variations on a basic theme, in adjoining counties: a Denbighshire wagon from Ruthin, believed to be of late-nineteenth-century date; and a Montgomeryshire wagon believed to have been built c.1800 at Llangadfan. (St Fagans: National History Museum)

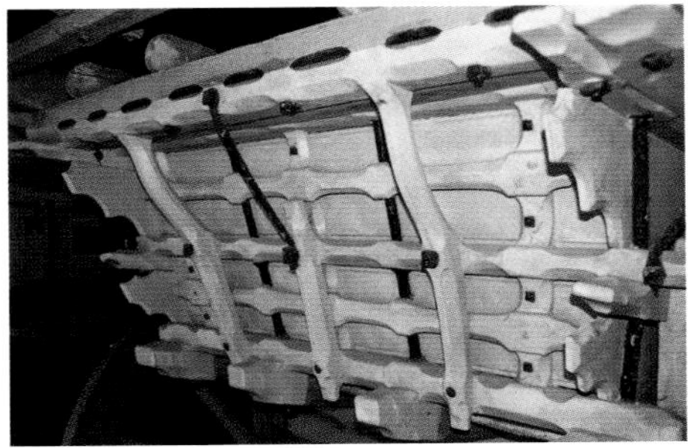

Box wagons from Sussex and Kent: (1) Sussex *c.*1870–80, (2) Sussex *c.*1910, (3) Kent – Staplehurst – *c.*1880, (4) Kent – Romney Marsh – *c.*1890–1900.

tells me it is now extinct'. An old broad-wheeler pulled by oxen would have been a memorable sight.

Kent and Sussex types are often confused, perhaps because of a geographical overlap in manufacture. But Kent is also unusual in that the undercarriage arrangement differs again, the pole being bolted directly to the central summer of the wagon bed, with no pole braces. Here strakes are less common, hoops being the fashion, and sides are panelled. The wagon body colour is buff and as usual everywhere the undercarriage has a salmon-pink to red colouring. Half locking is the norm.

A Sussex broad-wheel wagon from Nuthurst, seen here in 1919 outside the Southwater smithy, now re-erected at the Weald & Downland Open Air Museum.

Fit for purpose: another Sussex broad-wheeled wagon in a lane at Plumpton, fully laden and fresh from the mud of the field. Tall harvest poles create increased capacity and support the bulk of the load.

Another south-eastern feature is the use of a single wooden bar of variable local design forming the hind-rail in lieu of a tailboard. This relates in part to the carriage of sacks rather than loose crops, a reminder that there can also be found a specific hop-wagon design, robust and open sided, used for the carriage of large hop sacks.

With regard to Surrey, a strong sense of place and time comes through in George Sturt's classic *The Wheelwright's Shop,* an autobiographical account

'Hop picking': homeward-bound Londoners on their way to the railway station. This postcard was sent from Faversham in Kent in September 1905.

'Bringing hops into oast houses', with a hop wagon or tug on the right.

covering the years 1884–91 (first published in 1923) when many of the wagons and carts illustrated here would have been in regular use. Surrey wagons tended to be quarter lock, with virtually straight beds, spindled with a single mid-rail; they conformed to the blue and red colour pattern, although brown and buff are known. One particular distinguishing feature was the elegant shaping of the strouters, as elsewhere a subtle combination of chamfering and strength. Surrey-type wagons could be found in the counties around Surrey as well – Sussex and Hampshire especially.

Moving westwards into Dorset, the box shape remains strong and locally distinctive but there are quite clear variations within Dorset itself, especially east to west. Size varies, and some of the smallest box wagons found anywhere can be linked to west Dorset makers, a directory of *c.*1835 describing them as 'a light good sort of vehicle'. Some of these are no more than 8 feet long; at quarter lock, the 2 or 3 inch wheels are sometimes set wide on the axle, thus creating a strange sense of proportion. This apart, these are elegant wagons, very much in tune with their landscape. A strong West Country feature is the builder's name and address and a date in often highly decorative fashion on the tailboard although care must be taken that this does indicate construction rather than a subsequent repainting task or a change of owner. Colours were yellow and blue across the county.

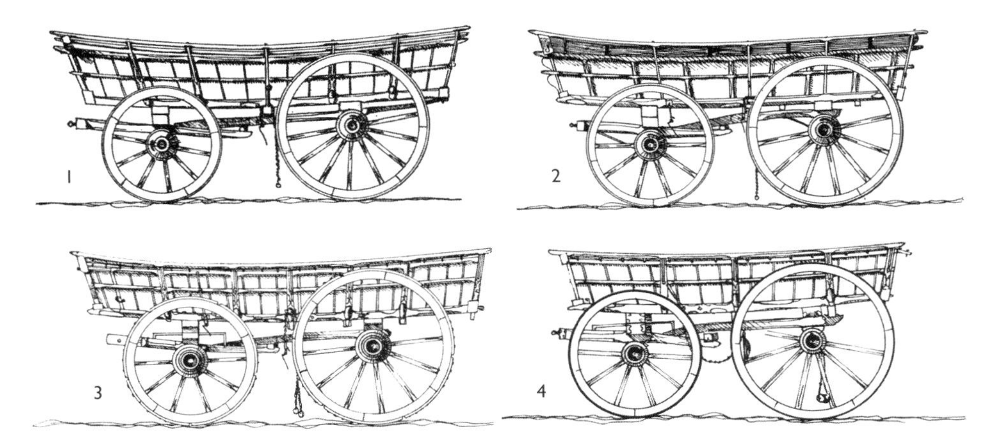

Above: Box wagons from central southern England: (1) Farnham, Surrey, c.1860, (2) south-west Wiltshire c.1900, (3) Dorset (Bridport area) c.1860, (4) Dorset c.1890–1910.

Left: Alf Squire, manor bailiff and head mason of Chideock in west Dorset, in his tiny donkey 'cart', a miniature Dorset box wagon.

James Arnold's delightful illustration of a Rossiter of Crewkerne Dorset box wagon of 1910.

T. COMELY & SONS
NOTGROVE STATION
GLOSTERSHIRE

BOW WAGONS

STUDYING the origins of wagon and cart design is rewarding but challenging; firm evidence is hard to find and there are many questions, not the least of which concerns the bow or hoop-raved wagon body, where the inner and outer raves rise in a gentle arch over the rear wheels. This shape, often elegant but sometimes rather contrived, can be found amongst long carts, where it is a simple reflection of the sweep over the wheels to protect the load. A similar shape is characteristic of the shallow and light Cornish wagons, which are otherwise little more than flat-bedded, shallow and splay-sided wagons. And the same profile can be found in various other harvest carts elsewhere around the country. It is without doubt an old as well as essentially practical design.

David Wray, whose line drawings illustrate this volume, suggested that in these Cornish types, both wagon and cart, lay the true origin of the hoop-raved wagon. The design then migrated up the country and found its best resonance, and indeed its most elegant form, in the many variations on wagon design which can be found in the old county of Gloucestershire and the counties immediately around. James Arnold suggested about a dozen distinct wagon styles in that county, where box and bow forms noticeably mingle.

All these ideas rely upon accepting that designs did migrate, as with the Dutch examples from the eastern counties already discussed, in a process where new ideas with beneficial economic effects mixed with the long-standing traditions and even dyed-in-the-wool habits of wagon and cart making, leading to change and further innovation.

There are many clues in support of such a theory to be found in the variations to both box and bow where they are used side by side, such as in Devon, Somerset and parts of Dorset and similarly across a large region extending from Bristol well into South Wales and up through the South Midland counties. Geraint Jenkins in his surveys coined the generic name of 'South Midlands bow wagon' as a reflection of this geographic spread. The curve creating the wheel arch varied, but essentially served to increase the load capacity whilst conveniently lowering the bed and maintaining stability.

Opposite: Outside the wheelwright's shop on the village green at Notgrove in the Cotswolds in 1937. A spindle-sided bow wagon stands in front of a box wagon, both locally made. The headboard includes an old spelling of the county – 'Glo'stershire'.

A bow wagon fitted with ladders or the tall corner poles used at harvest time could thus carry loads of several tons.

The type example is the 'Woodstock wagon', so called after Arthur Young's drawing of 1809. It has also been commonly called an 'Oxford' or 'Cotswold' type. William Marshall, in his *Rural Economy of Gloucestershire* published in 1796, had a telling phrase, describing this same type as 'beyond every doubt, the best farm waggon I have seen in the kingdom. I know not a district which might not profit by its introduction.' He noted its cost depending on size at £20 to £25 and weight of 15 cwt to 1 ton.

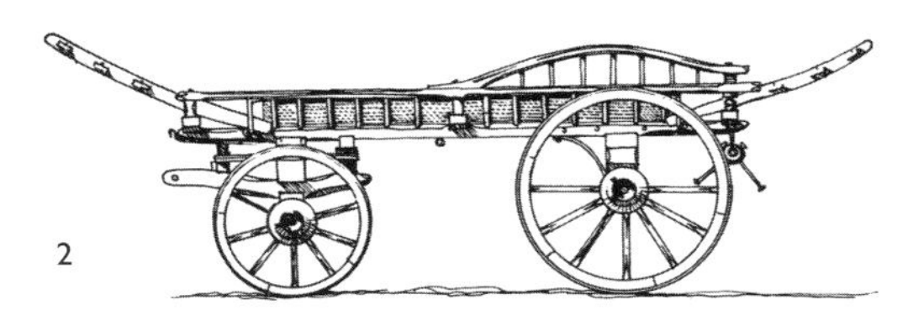

Above: (1) North Cornwall *c.*1890; (2) South Cornwall *c.*1910.

Above: Arthur Young's drawing of a Woodstock wagon, here published in *The Farmer's Companion* (1813).

Bow wagons from the South Midlands and Wessex:
(1) Oxfordshire or South Midlands type
(2) Wiltshire
(3) South Gloucestershire and Lower Severn

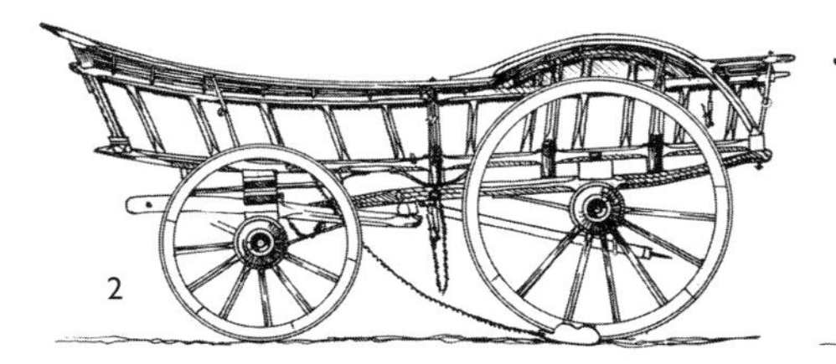

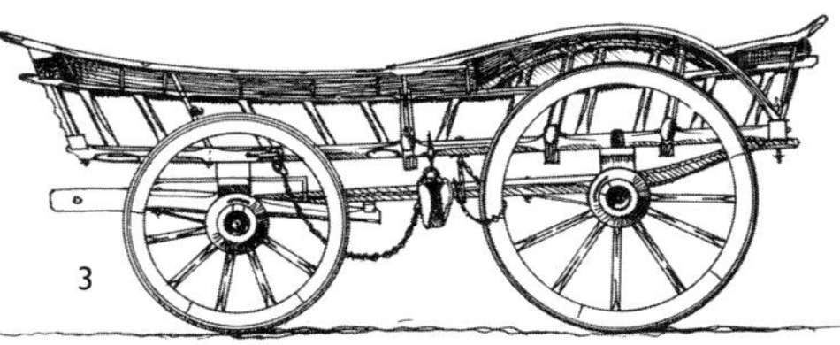

So the Oxfordshire (with the Buckinghamshire) style has endured as a symbol of elegant form suiting basic function. It consists of close spindles, a shallow bed and a large frontboard, carrying the owner's name, in a distinctive concave sweep to meet the outraves as they too rise up to the front. Iron brackets in an H or an N shape support the outraves and, for a 'true Cotswold', yellow and salmon pink was the usual colour combination.

Wiltshire and West Berkshire bow wagons have their own pedigree, with a blue body as standard. The definitive feature was the sweep of the hoop rave coming right down at the back of the wagon to join the rear or hind shutlock. Side lades were also boarded, not open as on the Oxford which tended to iron spindles instead. Other variants included straight sides (Wiltshire) or half locking (Berkshire), and the use of strakes to some examples of wider wheels.

The wheelwright's shop at Stratton near Bude, Cornwall, *c.*1910. The small cart on the left has a simple rave over the wheels.

A fine example of a Wiltshire/West Berkshire type of bow wagon, belonging to Ampney Mill at Ampney Crucis, outside the Wagon and Horses Inn, Cirencester, in April 1913. The inn was a regular stopping point for wagoners and carriers.

A focus of the bow-wagon form has been identified around the Lower Severn and the Vale of Berkeley, where the characteristic wagon is very low bodied, often with broad wheels, hooped as well as straked, and with the same sense of the sweep of the low body over the rear wheels, the rave coming down once again to the rear shutlock. A fine example from Dymock is also a reminder of another traditional use of the 'best' wagon: to convey the deceased owner on the last journey to the churchyard and cemetery, in this case in 1932.

The wagons of Glamorgan are related in virtually all details, similarly panel sided with a central mid-rail and the same sweep up to the front, creating a cow's-horn effect with the high outer raves curving upwards. A fine touch is the curved arch of open spindles crowning the top of the headboard, and it is tempting to see here, amongst the two-wheel wains of South and West Wales, another source of inspiration for the elegant bow design.

A final group in this broad classification of bow wagons is also distinguishable, by shape as well as by distribution. These wagons are particularly associated with Devon and Somerset but also found in Dorset, and are known as 'ship' or 'cock-rave' wagons. The names reflect the rising up of the body over the rear wheels, as in the shape of a ship or of a cockerel's tail. Instead of returning downwards to the shutlock, the hoop rave remains

Made in Bristol c.1880, this Vale of Berkeley wagon with ladders in position has the typical features of its type, including a half-moon front headboard with owner's name: George Cobb, Hill Ash Farm, Dymock. (Museum of English Rural Life, Reading)

in a horizontal plane behind the wheel. There are both panel- and spindle-sided wagons of this type, which although narrow are not waisted. There is good use of wooden strouters and chamfering, accentuated to good effect by being painted, and a notable use of twisting to the iron side-staffs by some makers. Somerset wagons tend to be larger than those in Devon or Dorset and blue is the standard body colour. As with other types of bow wagons, front and rear panels can be elaborately decorated.

The distinctive front end of a Glamorgan wagon, dated 1885. (St Fagans: National History Museum)

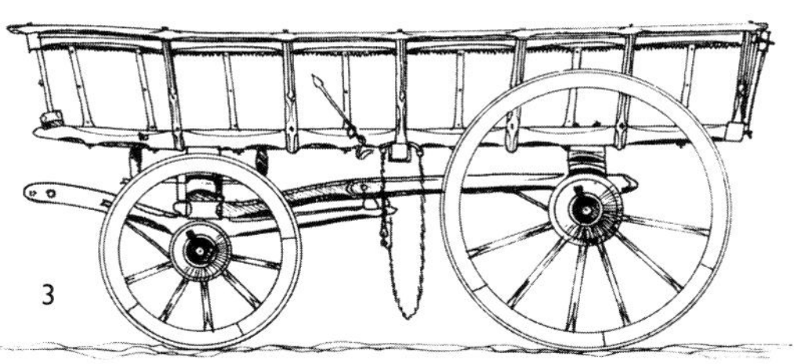

Bow and box wagons from south-west England:
(1) Somerset 'ship' or 'cock-rave' wagon,
(2) Devon 'ship' or 'cock-rave' wagon 1863,
(3) Devon 'buckrowe', 1892.

BOATS, BARGES AND TROLLEYS

Usually factory made, 'boat' and 'barge' wagons are distinctive variations upon older forms. Sometimes this is literally true as the new form of body was placed on a surviving older undercarriage, until that too came to be replaced. Economy of effort, of production methods and of cost was dominant. Uniform techniques were used, such as the use of bolts rather than mortise and tenon for fixing, although this is not to say that traditional builders did not use such methods too.

More standardised techniques particularly in making wagon bodies created a more utilitarian, less detailed and to many eyes less sophisticated vehicle, plain and simple and essentially functional – the typical barge wagon. Some of the artistic flourish disappears too, although that was always a luxury with farm wagons and carts, being working vehicles, earning their keep and little more.

Like so much of wagon and cart history, this transition from the products of village workshops into larger concerns, using recognisable factory methods of production, is not a measured sequence in terms of date nor consistent around the country. There is much overlap, even down to the final years of horse-drawn vehicle production in the twentieth century.

YORKSHIRE

The wagons of Yorkshire epitomise these factors, bringing together essentially utilitarian designs, whilst some local workshops established almost national reputations, seeming to bridge the gap into the world of plank- and panel-sided wagons. Firms such as Crosskills of Beverley in East Yorkshire were especially known for making wagons associated with that part of the county. William Crosskill, born in Beverley in 1799, had established his business by the time he was twenty-six and by the 1850s his wagons were well known in and beyond Yorkshire. The firm made its own iron axles and was an early user of cast-iron hubs for wagons; it also exported machine-made wheels with iron hubs across the Atlantic as early as 1853. Sissons of Beswick was another well-known maker, producing some fine spindle-fronted wagons.

Opposite: Fine spindle detailing decorates the low sides of this special-purpose cider wagon, perfectly described in its West Country setting in this postcard image by G. Boyns.

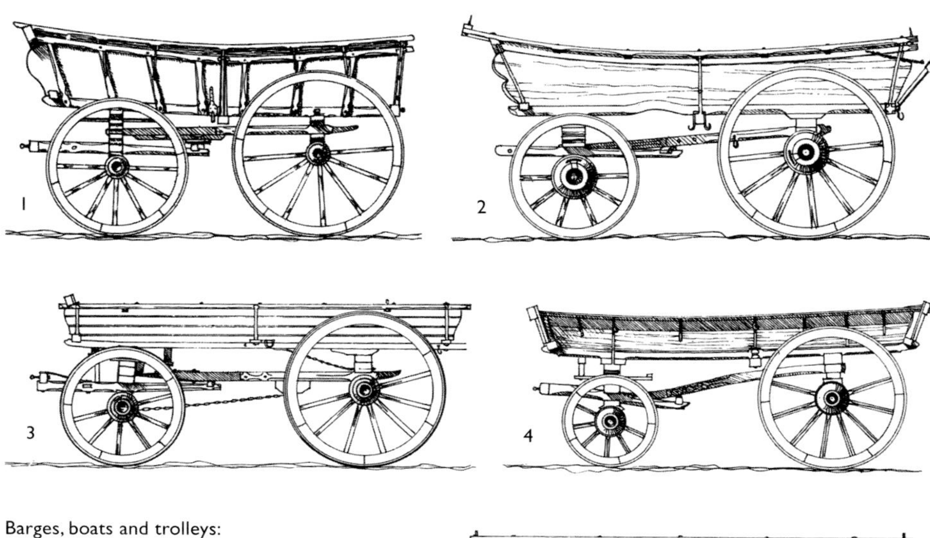

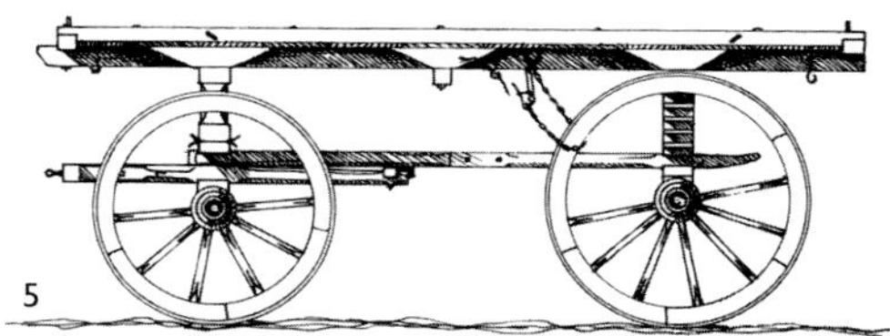

Barges, boats and trolleys:
(1) barge wagon by Crosskills, Beverley, Yorkshire, *c.*1910
(2) barge wagon body on an old bed and undercarriage, probably from Cambridgeshire
(3) West Midlands barge wagon
(4) three-quarter-lock boat wagon *c.*1900
(5) Midlands trolley *c.*1890–1900

Three different wagon types have been identified in Yorkshire, although overlap inevitably exists between them. The smallest is the Dales wagon, some 8 feet in length and perhaps only 3 feet 6 inches wide. Associated with the north-eastern Dales, they are quarter lock and panel sided. The Moors wagon is larger, typically up to 10 feet long, but with many similarities. It is the Wolds wagon which is the best known, and the largest. It is also called a

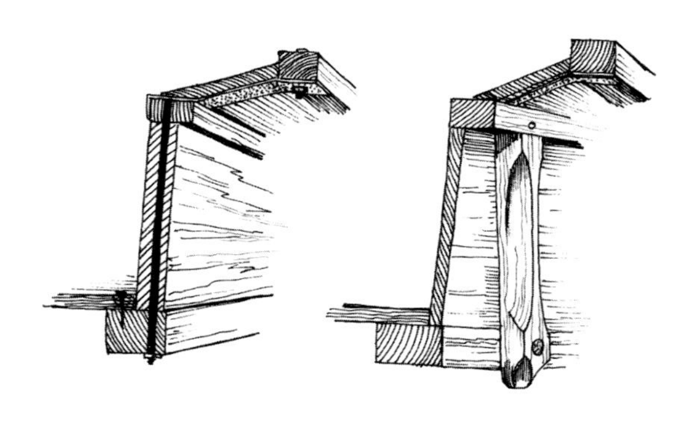

Economy and uniformity in construction methods: (*left*) in plank side construction, a single bolt locks together the top rave, the deal side board 2 inches thick, and the main side of the cart or wagon; (*right*) standards could be bolted to the main sides instead of mortised into them.

Barge-type double-shafted wagon by Woods & Co. of Stowmarket, dated 1894, originally varnished and subsequently painted. When new this wagon was a wedding present to the bridegroom from his future father-in-law. (Museum of East Anglian Life, Stowmarket)

Gloucestershire wagon made by Healey of Gloucester in 1924.

A plank-sided Yorkshire wagon, in typically light or stone colouring. The fore-carriage is designed around the use of a single pole.

A Yorkshire pole wagon (and another behind) forms part of a local celebration.

'pole' wagon, revealing one particular Yorkshire trait – the use of a single pole rather than shafts. The fore-carriage is specially designed to house the pole whilst still allowing for shafts, and the whole is reminiscent of ox-drawn rather than horse-drawn wagons, a trait shared with Sussex into the twentieth century. Other features are the cross-boarding of the bed of the wagon, and the continuing use of wooden axles, which explains the surviving tradition of large wooden naves.

BOAT AND BARGE WAGONS

Boat wagons are named after their shape, their side boards sloping outwards like a boat's hull and often made with two planks, the upper serving as the lade. The whole body is shallower than that of other wagons so far noted, and in some cases there is little difference between boat wagons and trolleys or flat-bed carts with removable side panels. Boat wagons usually had a three-quarter lock, which made for greater flexibility and, being smaller all round, offered much greater manoeuvrability than the old-style quarter-lock wagons. They could not necessarily carry the same-size loads.

Some well-known wagon makers' names are associated with this type, Taskers of Andover in Hampshire and E. & H. Roberts of Deanshanger in Buckinghamshire being two of the best known. Nor were boat wagons only factory made, as many local wheelwrights' shops also produced them.

Both barge and boat types often had iron wheel-centres to replace the old elm naves, the maker's name forming part of the detailing. The name of David Ward's works at Long Melford in Suffolk is often seen on restored

Top left: The West Country tradition of painting (and here restoring) the builder's details on the tailboard shows well in this example from the Wiltshire/Dorset border.

Top right: Typical boat wagon design, still with a very gentle sweep to the front. This restored example dates from 1929 and was made at the Birdbush Wagon Works of W. H. Harrington & Son at Donhead St Mary near Shaftesbury, Dorset.

wagons, as is that of Tasker. Although many surviving examples are of twentieth-century date, the use of cast-iron wheels dates from before 1850, a reminder of the overlap between the use of iron and wood.

TROLLEYS

Most of these makers are also associated with the production of trolleys and other forms of flat-bed wagon. Names vary; for example, a 'rulley' is a northern counties' name. With the trolley essentially a platform on wheels, the variations came as much in the undercarriage as the bed, including in the height of the undercarriage from the ground. The addition of layers of timber, or blocking, served to raise the bed while creating a solid working platform and plenty of manoeuvrability underneath. Side frames of various depths could be added, to create a box effect. Sometimes a low, curved, front headboard, painted with the owner's name, completed the design.

Trolleys were mass produced and in common use by a wide variety of businesses and tradespeople, as well as farmers and on the land. With the addition of braking systems and the introduction of iron springing to the

Left: Enjoying the ride at harvest time. Trolleys of this kind could later be converted with pneumatic tyres and tractor bars.

A West Midlands trolley from Presteigne in Radnorshire, the platform noticeably higher off the ground than on other types. Its owner called it 'a singularly good example, the wheels being big enough to let the deck clear the gateposts of the usual 9 foot gate'.

axles, they were multi-purpose vehicles familiar in streets for bulk deliveries such as coal.

As specialist demands grew, so more robust forms of trolley were manufactured – for the railways for instance and in support of steam road-vehicles. Some of the major makers included firms like Bristol Wagon & Carriage Works Company. Their catalogues are immensely useful records of the variations on offer. From there it was a short step to the lorry and thence down to the present day, with vehicles petrol – and then diesel – driven rather than horse powered, on pneumatics instead of wooden wheels hooped in iron, and increasingly standardised across the country. But their origins lay long before, in the local workshops of wheelwright and blacksmith.

A sprung flat-bed coal wagon with detachable sides and handbrake, made by Lloyd & Sons of Ludlow, Shropshire, and owned and used locally. On the right, a sprung Jersey box wagon, the wheel hubs all reading 'Lane/Jersey'.

AFTERWORD

In compiling this book, the work of various writers and researchers in drawing and recording wagons and carts has been much admired and drawn upon. Each struggled with matters of interpretation whilst constantly adding to their knowledge: it is fascinating to compare the assessments each made. They had in common a commitment to the subject, while being well aware of the significance of regional differences. All noted that in the wagons and carts produced around the country from the mid eighteenth to the mid twentieth centuries there was a strong sense of regional identity and regional ways of life. With this came a distinctiveness in both style and detail which remains the hallmark of the vehicles. This same distinctiveness was found in the people who made, owned, repaired and cared for the wagons and carts. Today we call it regional diversity. It has a long history.

Lined up in the sunshine awaiting sale day: a spindle-sided bow wagon, an Essex wagon from Williams of Colchester, a Cook of Lincoln wagon dated 1910, and a South Lincolnshire spindle-sided wagon (Hankelow, Cheshire, 2006).

GLOSSARY

axle arm – the iron or wooden spindle upon which the wheel is turned *(h, 4)*
axle bed – the wooden beam to which the axle arm is fitted; also called an 'ex' bed
axle tree – an all-wood axle with wooden arms
bed – the timbers forming the base of the body of wagon or cart
bolster – wooden crosspiece above the axle *(g and o)*
box – the hardened metal centre of the nave which runs on the axle
braces – a pair of timbers attached to the pole to keep it at right angles to the rear axle *(i)*
chamfering – the process of shaving the timbers in order to reduce weight, resulting in a decorative feature
collet – a washer on the end of the axle arm to protect the lynch pin *(5)*
copse – an iron stay to keep the outrave in position
coupling pole – the long timber pole joining the fore-carriage to the rear
crook – the curved separate section of the front sides of a waisted wagon *(m)*
crossledge – the central cross member of a wagon body *(f)*
dish – the concave arrangement of the spokes of a wheel and the sloping angle of the wheel itself away from the wagon or cart body
dog stick – a wooden stick with a forked metal end fixed to an axle tree; dragging on the ground, it acted as a brake

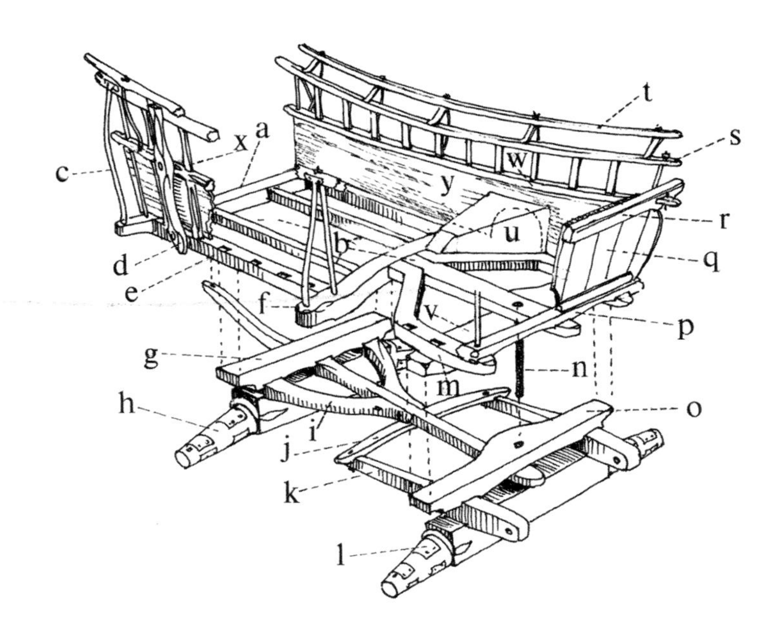

Exploded diagram of a Suffolk wagon (for key, see glossary items).

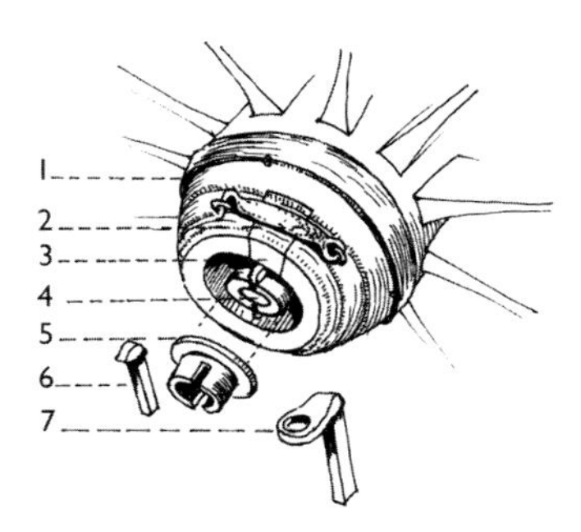

A typical nave, hub or stock with fittings for an iron axle arm (for key, see glossary items).

drag or **drug shoe** – a cast-iron wedge-shaped shoe to act as a brake; also called a 'bat' or a 'skid pan'

draught pin – a metal pin used to fix shafts to the hounds or splinter bar

felloe – a section of the rim of a wheel (pronounced 'felly')

fifth wheel – flat rings of iron usually about 2 feet in diameter, fixed to the turning pillar and bolster of a fore-carriage in order to maintain its vertical alignment

forehead – the upper part of the front end of a wagon or cart body (*r*)

front board – the front end of a wagon or cart body (*q*)

headboard – front board

hermaphrodite – a cart body with an added or detachable fore-carriage to increase carrying capacity, i.e. half cart, half wagon

hind – the rear of a vehicle, hence hind wheel, hind shutlock etc.

hoop – the iron banding which holds a wooden wheel together

hoop-raved – the shape of a wagon body which rises over the rear wheel in the form of an arch

hounds – parts of the fore-carriage framework (*k*)

iron wearing plates, clouts or **cletes** – iron plates fixed to a wagon or cart for protection, e.g. on an axle or where the fore-wheel meets the body when turning (*l*)

keys – timbers placed at right angles to the summers to provide fixing points for long boards

king pin – the long pin passing through pillow, bolster and axle providing the pivot for the fore-carriage (*n*)

ladders – fore and rear detachable wooden frames to extend the carrying capacity of a cart or wagon

lade – overhanging shelf projecting from the top rave of the side boards to increase carrying capacity and prevent the load fouling the wheels

locking arch – the cavity formed in the bed of a half-locking wagon to allow the fore-wheels maximum turning space (*u*)

locking chain – a long and a short length of chain fixed to the sides of a wagon or cart to lock a wheel and prevent it turning

long boards – boards running parallel to the cart or wagon sides, as distinct from cross boards, which are fixed directly to the summers

lynch pin – the pin at the end of the axle arm to keep the wheel in place (6, 7)

mid-rail – the horizontal timber dividing the wagon sides; also called a 'middle rave'

nave – the elm hub into which the wheel spokes are mortised

nosepiece – the front crosspiece of a wagon or cart bed (*p*)

outrave – the outer rail which overhangs the side of a wagon or cart, held in position by the copses (*t*)
pillow – cross member at the front of the body bearing on the bolster of the fore-carriage (*v*)
propstick – a short stick hung below the shafts of a cart; when dropped down with the cart at rest it took some weight from the horse
pummel – the rear extension of the bed of a dung cart to take the bump on tipping
rails – the cross members on a headboard or tailboard (see also **forehead**)
rave – longitudinal timber attached to the sides of a wagon, such as the top rave (*s*), the outrave (*t*) or the middle rave (mid-rail) (*w*)
roller scotch – a form of brake: a cylindrical piece of wood for use with the rear wheel to 'scotch' any rearward movement going uphill; fixed with a roller chain
shutlock – the end cross member of a wagon or cart body (*a*)
shutters – cross timbers joining a pair of shafts, hounds or the main sides of a dung cart
side – the outer lower timbers of wagon or cart bed (*e*); also called the 'sole'
side boards – the body of the wagon or cart above the bed (*y*)
slider – the slightly curved timber connecting the rear end of the hounds (*j*); also called the 'sway bar'
sole – see **side**
spindle – wooden upright fixed to the wagon or cart sides to provide support and/or as part of the design
splinter bar – the cross piece on the fore-carriage to which the shafts are fixed
staff or **staves** – forged iron brackets which buttress the side boards against the weight pushing outwards, such as the hind staff (*c*)
standards – the slats fixed to the body planks to resemble panels in 'panel-sided' vehicles (*x*)
stock – an alternative name for the nave; 'stock bonds' (*1*) are iron rings shrunk on in front and behind the spokes to prevent them splitting
stopper – a wooden block closing the slot cut in the face of the nave or stock to allow the lynch pin to be withdrawn (*3*), held in place by a 'stopper clasp' (*2*)
strake – an iron tyre made in sections and nailed to the wheel
strouter – a wooden support for strengthening the wagon side, often elegantly shaped (*d*)
summers – part(s) of the body framework of a wagon or cart, parallel to the sides and joined to the shutlocks front and rear (*b*)
sway bar – see **slider**
tip stick – a means of controlling the degree of tip on a cart body

FURTHER READING

Arnold, James. *The Farm Waggons of England and Wales*. John Baker, 1969.

– *Farm Wagons and Carts*. David & Charles, 1977.

Beach, Bob. A series of studies in *Heavy Horse World* magazine of regional types, including Northumberland (Summer and Autumn 1993), Bow wagons (Summer 1994), East Anglia (Winter 1994) and Yorkshire (Spring 1995).

Hennell, Thomas. *Change in the Farm*. Cambridge University Press, 1934.

Jenkins, J. Geraint. *Agricultural Transport in Wales*. National Museum of Wales, 1962.

Jenkins, J. Geraint. *The English Farm Wagon*. David & Charles, 3rd revised edition, 1981, of a study originally published by the University of Reading with Oakwood Press, 1961.

Harris, Stephen. *Farm Wagons of East Anglia*. Boydell Press, 1979.

Mills, Dennis. Two studies on hermaphrodites in East Anglia and the East Midlands published in *Heavy Horse World* magazine, Spring and Summer 2007 and *Folk Life*, Vol 46, 2007–8.

Smith, David. *A Dictionary of Horse Drawn Vehicles*. J. A. Allen & Co, 1988.

Sturt, George (George Bourne). *The Wheelwright's Shop*. Cambridge University Press, 1923 and 1963.

Thompson, John. *Horse-drawn Farm Vehicles*, no. 8 in his Sourcebook series. 1980.

– *Carts, Carriages & Caravans*. Self-published, 1980.

Vince, John. *Discovering Carts and Wagons*. Shire Publications, 1970 (and two albums in the Shire Modelmakers series: *Modelling Farm Wagons* and *Modelling Farm Carts*, both 1977).

– *An Illustrated History of Carts & Wagons*. Spurbooks, 1975.

Viner, David. 'James Arnold – the man who knew, drew and loved waggons' in *Heavy Horse World*, Winter 1999.

– A series of studies in *Heavy Horse World* magazine of museum and private collections containing traditional farm wagons and carts from around the country; see especially Tiverton Museum of Mid-Devon Life (Autumn 1999), Cotswold Heritage Centre at Northleach (Autumn 2000), Weald & Downland Open Air Museum (Spring 2001), the Waggons of Severn, Usk and Wye (Summer 2005), and the collection at Dyrham Park (Summer 2007). See Winter 1998, Winter 2006 and Summer 2008 issues for dispersal sales of important private collections, and Winter 2007 for the Birdbush Wagon Works of Harrington & Sons, Donhead St Mary.

Heavy Horse World magazine is published quarterly from Lindford Cottage, Church Lane, Cocking, Midhurst, West Sussex GU29 0HW. Telephone: 01730 812419. Website: www.heavyhorseworld.co.uk

WHERE TO SEE WAGONS AND CARTS

Restored wagons and carts can often be seen displayed at shire horse centres, county and local shows, and at some steam rallies and farm open days. Private collections are by their nature not usually accessible, except by invitation.

The best way to see and enjoy the range of preserved (and often unrestored) agricultural types is on display in museums, particularly at the various regional collections established around the country during the past half century. Sites marked * also have significant items held in reserve store. Single vehicles on display or in store are omitted.

The Guild of Model Wheelwrights holds displays at various venues around the country – see www.guildofmodelwheelwrights.org for details.

ENGLAND

BERKSHIRE

Museum of English Rural Life at the University of Reading, Redlands Road RG1 5EX.
Telephone: 0118 378 8660. Website: www.merl.org.uk
A collection designated of national importance.

CAMBRIDGESHIRE

Wimpole Home Farm (National Trust).
Telephone: 01223 206000. Website: www.wimpole.org

CORNWALL

Shire Horse Farm and Carriage Museum, Treskillard, Redruth TR16 6LA.
Telephone: 01209 713606

DEVON

Museum of Mid Devon Life at Tiverton EX16 6PJ.
Telephone: 01884 256295. Website: www.tivertonmuseum.org.uk

Bicton Park Countryside Museum, East Budleigh EX9 7JT.
Telephone: 01395 568465. Website: www.bictongardens.co.uk

CO. DURHAM

*Beamish, North of England Open Air Museum, Stanley DH9 0RG.
Telephone: 0191 370 4000. Website: www.beamish.org.uk.

GLOUCESTERSHIRE

*Northleach (formerly the Cotswold Heritage Centre). Collections including wagons in store; access by arrangement with Corinium Museum, Cirencester.
Telephone: 01285 655611.

A collection of twenty-three model wagons made by H. R. Waiting during the period 1931–8 for the owner Mr Charles Wade forms part of the display at Snowshill Manor (National Trust). Telephone: 01386 852410. Website: www.nationaltrust.org.uk

HAMPSHIRE

*The county's collections are stored at Chilcomb House, Winchester SO23 8RD; access by arrangement. Telephone: 0845 603 5635. Website: www.hants.gov.uk/museums

KENT

Museum of Kent Life, Cobtree, near Maidstone ME14 3AU. Telephone: 01622 763936. Website: www.museum-kentlife.co.uk

Brook Agricultural Museum, near Ashford TN25 5PF. Telephone: 01304 824969. Website: www.agriculturalmuseumbrook.org.uk

LINCOLNSHIRE

*Museum of Lincolnshire Life, Burton Road, Lincoln LN1 3LY. Telephone: 01522 528448. Website: www.lincolnshire.gov.uk/museumoflincolnshirelife

Normanby Hall, near Scunthorpe DN15 9HU. Telephone: 01724 720588. Website: www.northlincs.gov.uk/museums

NORFOLK

Banham Zoo. Telephone: 01953 887771. Website: www.banhamzoo.co.uk

*Gressenhall Farm and Workhouse (Norfolk Rural Life Museum), near Dereham. Telephone: 01362 860563. Website: www.museums.norfolk.gov.uk

Norfolk Shire Horse Centre, West Runton, near Cromer NR27 9QH. Telephone: 01263 837339. Website: www.norfolk-shirehorse-centre.co.uk

OXFORDSHIRE

The county's collections are displayed at: Cogges Manor Farm, Witney OX28 3LA (Telephone: 01993 772602. Website: www.cogges.org); Swalcliffe Barn, near Banbury OX15 5DR. Telephone: 01295 788278; and in the *Resource Centre at Standlake (Telephone: 01864 300972. Website: www.oxfordshire-collections.org.uk)

RUTLAND

*County Museum, Oakham LE15 6HW. Telephone: 01572 758440. Website: www.rutland.gov.uk/museum

SHROPSHIRE

Acton Scott Historic Working Farm, Church Stretton. Telephone: 01694 781306. Website: www.actonscott.com

SOMERSET

*Rural Life Museum, Abbey Farm, Glastonbury BA6 8DB.
Telephone: 01458 831197. Website: www.somerset.gov.uk/museums

STAFFORDSHIRE

*County Museum, Shugborough ST17 0XB.
Telephone: 01889 881388. Website: www.shugborough.org.uk

SUFFOLK

*Museum of East Anglian Life, Stowmarket IP4 1DL.
Telephone: 01449 612229. Website: www.eastanglianlife.org.uk

SURREY

Old Kiln Museum, Tilford, near Farnham GU10 2DL.
Telephone: 01252 795571. Website: www.rural-life.org.uk

WARWICKSHIRE

Mary Arden's House, Wilmcote, Stratford upon Avon CU37 9VN.
Telephone: 01789 293455. Website: www.shakespeare.org.uk

WEST SUSSEX

*Weald & Downland Open Air Museum, Singleton, near Chichester PO18 0EU.
Telephone: 01243 811363. Website: www.wealddown.co.uk.
A collection designated of national importance

WILTSHIRE

Lackham Museum of Agriculture and Rural Life, near Chippenham SN15 2NY.
Telephone: 01249 466800. Website: www.lackhamcountrypark.co.uk
*Science Museum outstation and store, Wroughton SN4 9NS– access on open days.
Telephone: 01793 846200. Website: www.sciencemuseum.org.uk

WORCESTERSHIRE

County Museum, Hartlebury Castle, Kidderminster DY11 7XZ.
Telephone: 01299 250416. Website: www.worcestershire.gov.uk/museum

YORKSHIRE

Shibden Hall, Halifax HX3 6XG.
Telephone: 01422 352246. Website: www.shibdenpark.com
Museum of Farming, Murton, York YO19 5UF.
Telephone: 01904 489966. Website: www.murtonpark.co.uk
*Castle Museum, York – all in store - YO1 9RY. Telephone: 01904 687687.
Website: www.yorkcastlemuseum.org.uk

Ryedale Folk Museum, Hutton le Hole YO62 6UA.
Telephone: 01751 417367. Website: www.ryedalefolkmuseum.co.uk

WALES

*St Fagans: National History Museum, Cardiff CF5 6XB.
Telephone: 029 2057 3500. Website: www.museumwales.ac.uk

Usk Rural Life Museum, Monmouthshire NP15 1AU.
Telephone: 01291 673777. Website: www.uskmuseum.org.uk

SCOTLAND

National Museum of Rural Life, Wester Kittochside G76 9HR.
Telephone: 0131 247 4377. Website: www.nms.ac.uk/rurallife

Highland Folk Museum, Kingussie and Newtonmore PH21 1JG.
Telephone: 01540 673551. Website: www.highlandfolk.com

ISLE OF MAN

*National Folk Museum, Cregneash IM9 5PT. Telephone: 01624 648000.
Website: www.gov.im/mnh

NORTHERN IRELAND

Ulster Folk and Transport Museum, Cultra, Holywood BT18 0EU.
Telephone: 02890 428428. Website: www.nmni.com

The general arrangement of a typical undercarriage, seen fore and rear.

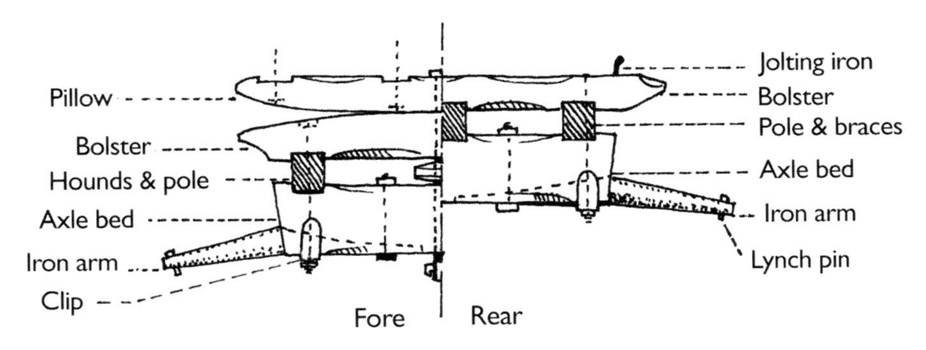

INDEX

Page numbers in italic refer to illustrations